# Occam's Universe

~~

## The End of an Error

@2020 Jack McNally

All rights reserved.

No part of this publication may be reproduced, stored in or introduced into a retrieval system, or transmitted, in any form or by any means (electronic, mechanical, photocopying, recording, or otherwise) without the prior written permission of the author. The scanning, uploading, and distribution of this book via the internet or via any other means without permission is illegal and punishable by law.

Please purchase only authorized electronic editions and do not participate in or encourage piracy of copyrighted materials. Your support of the author's rights is appreciated.

Cover image courtesy of NASA Marshall Space Flight Center

# PREFACE

Before anything can change or be changed, act or be acted upon, it must exist.

Sounds like a no-brainer; right?

Think again. It is probably the most profound observation you will ever contemplate.

Why? If existence is required in order for change to occur, then cause-and-effect is a "function of" the phenomenon of existence. It means existence is the source of cause-and-effect, NOT the result of it.

This isn't exactly rocket science, it is simple, basic logic. The purpose of science is to understand the nature of the cosmos. When their theories don't make sense, they are nonsense. Why do disciples of "Creation" (aka cosmic expansion) assume the cart precedes the horse?

The phenomenon of existence is reconciled with the canons of logic by a principle, not a process.

# Occam's Universe

**Creation Ex Nihilo:**

Conventional wisdom has concluded the universe must have come from somewhere, and the idea it was ushered into being by some primordial nascent event appeals seductively to human intuition. The very process of thought is governed by the laws of cause-and-effect, so ever since scholars began to reason, it has been instinctively presumed the physical presence of the cosmos, itself, must have been the result of a spontaneous feat of creation.

The existence of nothing is considered the only natural phenomenon that requires no rational justification. Since the constraints of logic don't apply to nothingness, most theories of cosmic origin pose a timeless, dimensionless nothingness in which space, time and the physical universe suddenly materialized in some miraculous omnificent event. But a strict adherence to the premise of causation would require whatever precipitated that event to have also been created, a descendant of an earlier predecessor similarly predated by a never ending procession of ancestry. The infinite "chicken or the egg" redundancy inherent to any cause-and-effect approach to the phenomenon of existence implies no logical beginning unless it includes some form of spontaneous source not derived from causation. But that would only invalidate the premise entirely. If such a source could exist without causation why couldn't everything else?

Something more fundamental than cause and effect justifies the phenomenon of existence.

In the 17th century, Isaac Newton proposed an infinite and eternal or "static" universe; infinite because the effect of gravity on a finite field of mass would eventually cause it to collapse into a single body of matter, and eternal because God was thought to be eternal. For somewhat more scientific reasons, Einstein initially agreed with the premise of an eternal cosmos, but when he applied his General Relativity calculations to gravity, his equations required a spatially finite universe. Soon thereafter, Edwin Hubble confirmed a red shift in elemental spectra from distant galaxies. Assuming it was a Doppler effect, he concluded an expanding universe must have had both a spatial and temporal point of origin called a "singularity".

Hubble's theory ultimately convinced Einstein to dismiss the premise of eternity, discard his cosmological constant, and embrace "Big Bang" as the mysterious event that gave birth to the universe.

In just a few short decades, Newton's infinite and eternal cosmos became both finite and temporal, born only 200 million years or so before our Milky Way and just ten billion years before our sun and all of its offspring emerged from the primordial cosmic soup.

Expansion and inflation theories both rely on Hubble's red shift being Doppler related. Before 1931, Einstein was skeptical; in part because gravity's drag on light was enough to explain the observation. Light traversing a gravity field loses energy. Its velocity must stay constant so a longer wavelength (red) results. This is called gravitational redshift. Doppler analysis suggests the more distant galaxies are moving away faster than the speed of light. That is sorely frowned upon in the realm of contemporary cosmology, but to the gravitational redshift fans, that substantially altered wavelength makes perfect sense. If you rest a white cue ball in a tall tube of cranberry juice, the deeper it sits the redder it appears The more distant the galaxy and the more gravity a wave encounters, the redder the shift. Forget stars, planets, moons, and asteroids, space isn't empty; it's populated by random particles. How many mass-laden obstacles would a photon encounter in a multi-billion light year trek?

But expansion theorists justify that excessive shift by adjusting each galaxy's Doppler calculated speed by a fudge-factor determined by some hypothetical rate of spatial expansion. Another fudge-factor derived from a theoretical bout of hyper-inflation during the earliest instant of "creation" ostensibly explains how a 14 billion year old universe has a 46 billion light year radius. That hyper-inflation supposedly allowed expansion faster than light.. Such circular reasoning is conjured into expansion theory to comply with the laws of nature.

Other serious incongruities plague the current cosmic model. Our Milky Way regularly collides with other galaxies. Andromeda is on the way. It should be here in about 3.7 billion years. One might suspect after only 14 billion years, galaxies expanding away from a common point of origin should likewise be moving away from each other; exceptions would be rare to non-existent. Luminosity studies of type 1A supernovae from distant galaxies suggest the cosmos isn't just expanding, it's actually accelerating, but comparative analyses of light versus cosmic microwave background radiation disagree on the rates of both expansion and acceleration. Expansion math breaks down in the face of relativistic exploration and relies too heavily on hypothetical dark energy and dark matter. Black holes twelve billion years old are too big to have formed only two billion years after creation and instead of containing just hydrogen and helium, stars born shortly after the "Big Bang" test positive for heavier elements that would have taken many more eons to evolve from scratch. This doesn't pass the "smell test", yet it has become the conventional wisdom and the basis for modern theoretical astronomy.

The mathematical flaws of a falsely premised model are easily reconciled by additional faulty hypotheses and new esoteric calculations reverse engineered to force the correct results. There is one very simple and irrefutable reason to reject every premise of cosmic creation. It's the basic, self-evident axiom: *Before a thing can act or be acted upon, change or be changed, it must exist.*

Creation is a myth.

~~

**Existence (Not Creation) Ex Nihilo:**

Something more fundamental than cause and effect justifies the phenomenon of existence. The realm of cause-and-effect is ruled by fundamental laws of nature called principles; governing dynamics that regulate the processes they engender. Existence is the source of all processes, so it is certainly more fundamental than cause-and-effect. If both existence and principles are more basic than processes, would it not be reasonable to assume the phenomenon of existence is logically justified by a principle instead of a process?

There exists just such a principle. It is, in fact, the most basic of all principles. It's the foundation upon which all other principles rely. It's universally accepted and used by all of us every day yet its true significance remains undiscovered, hidden in plain sight.

It's not by mere coincidence mathematics encodes its logic into a device called an equation which requires its elements to be equivalent on opposite sides of the argument. Newton's third law codified this natural balance by stipulating every action precipitates an equal and opposite reaction.

We customarily define things within the context of three basic parameters: quantitative (how much), spatial (where) and qualitative (what). Every numeric value and each spatial vector in the polar spectrum has an opposite equivalent. Shouldn't qualitative values also have opposite equivalents?

The Standard Model of Particle Physics (SMoPP) tells us our universe is composed of irreducible, structureless particles and antiparticles. If everything in the cosmos was spawned at some spatial and temporal point of singularity, it might be reasonable to expect that such an "Ex Nihilo" process would produce separate countervailing existences. The equivalent of nothing would still exist and the cosmic architecture would be in equilibrium. But SMoPP admits there's a whole lot more matter than antimatter floating around the cosmos and they seem to lack any rational explanation for it. When what particle physicists call "matter and antimatter" meet in a laboratory, mass disappears and energy is released. Einstein tells us mass and energy are just the same critter in a different coat. If two truly opposite existences ever met they wouldn't release energy. No mass or energy would remain; they'd simply cease to exist. What physicists like to call "annihilation" is merely conversion.

    To paraphrase an ancient Indian parable, when it comes to particles smaller than an atom, the interpretations of contemporary particle physicists are much like those of six blind scientists and the elephant. Subjective perspectives invite speculative conclusions and it appears they haven't really found any truly opposite existences. Their particles and antiparticles are simply elements in opposing condition. They readily admit they can't verify they're irreducible and it's entirely possible they wouldn't recognize a fundamental particle even if they isolated one. During collision experiments, the swirling traces CERN detects may not represent true particles. They are most likely signals of energy propagation that can be detected – but not captured.

The SMoPP does espouse the principle of qualitative balance somewhat, but not strictly. Big Bangers suggest more matter than antimatter was created "in the beginning", so whatever mutual annihilation that might have occurred still left enough stuff around to form a (finite) universe. It's a weak excuse for an unbalanced equation. There was no "beginning" and "matter and antimatter" aren't opposite existences. If they're wrong so far, why should we believe them when they tell us irreducible particles are structureless? What if qualitative values and their opposites are all nestled comfortably within the framework of every irreducible particle?

~~

**An Instance of Null Value:**

We could never dissect an irreducible particle and then examine its parts; it's inseparable, made only of itself. That's why it's called irreducible. Such a particle can't have independent sub-elements but may well have a substructure of inter-dependent equal and opposite properties. A truly fundamental particle would exist "Ex Nihilo" as an integrated instance of null qualitative value.

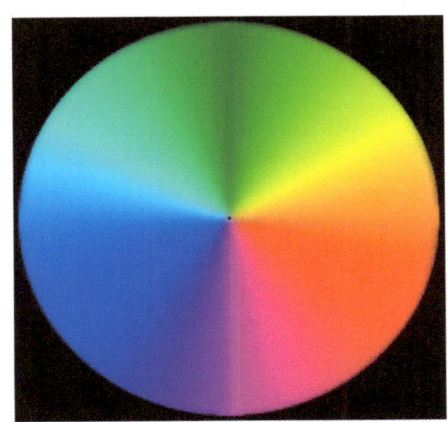

Let us assume BLACK represents a null qualitative value. Within the realm of color, the opposite of MAGENTA is GREEN. Equal proportions of those two colors produce BLACK. But GREEN is itself a blend of two colors, CYAN and YELLOW.

Just as the quantitative value of zero is equal to two opposing numbers +1 and -1, the qualitative offsets of BLACK are comprised of three opposing colors MAGENTA, CYAN and YELLOW, each in precisely equal proportion. If just one pixel of this color wheel is removed, the remaining pixels would no longer have a null value (BLACK) and the law of natural balance would be broken. Every point in the particle is a co-dependent instance of sub-quality.

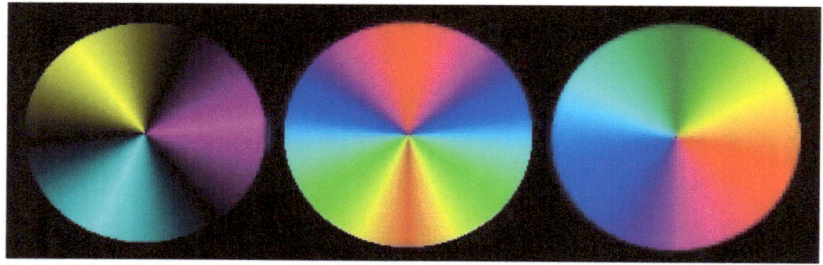

The images above show three iterations of the same fundamental particle. All of the subqualities are still present, just arranged into different configurations. There's a basic law of physics that states two things can't simultaneously occupy the same space (this is why we have automobile insurance), but two subqualities in the field of an irreducible fundamental particle are not "two things", they are mutually co-dependent instances of the same element. The rules of conduct within an entity may be vastly different from those that govern the interaction between independent entities. Its subqualities may be able to combine and separate to produce a range of altered states of being. They may come in an infinite variety of flavors.

Fundamental particles aren't composed of independent components, so no portion of an entity could be separated from the remainder. When a composite is severed, as in slicing a loaf of bread or tearing a piece of paper, forces holding independent particles adjacent to each other are overcome by the force used to separate the material. But the field of existence within an entity is continuous. In order to sever an entity, something must be inserted between two of its continuous points. Two independent existences cannot occupy the same space, so the point you are attempting to cleave would simply move. To sever an entity at any point within its domain would require the point of separation to physically cease to exist. Change requires existence, so any hypothetical "point of annihilation" would be a limit to its ability to change or be changed.

So, we have an entity. It's pretty. It has equivalent qualities and anti-qualities, but it also has size and a location in space. It must be SOMETHING, so it MUST have been created; right?

Well, from a relative perspective, this model of an irreducible entity is only qualitatively equivalent to nothing, but from an infinite or the "cosmic" perspective, the values of all criteria we use to define a physical existence (qualitative, quantitative and spatial) are rendered absolutely null.

Using any given point in space as the origin for a Euclidean X-Y-Z axis, one may extend theoretically endless (thus equidistant) lines throughout the infinite spectrum of the X-Y-Z coordinates. This inscribes a sphere that hypothetically encompasses the universe. Since the same holds true for any given point, every point in the universe must be its center.

Per the above, each point in the cosmos is the center of the universe. The size of any body (or any finite size) compared to infinity has no relative value and every qualitative value is offset by an opposite equivalent. From an infinite or "cosmic" perspective, our model ultimately has no qualitative, quantitative or spatial values.

That unique, intrinsically logical primordial substance we commonly call "Nothing" is only our abstract interpretation of the countervalence that engenders the cosmos. Opposing symmetry has always existed. It is the common essence of every element in the cosmic spectrum and the fulcrum of an eternally balanced perpetual system.

~~

**Balls and saddles:**

Why do so many contemporary cosmologists believe in a finite universe?

If a temporal universe had been born from a singularity, unless it expanded at an infinite rate or for an infinite period of time it would have to be finite. It fits Big Bang's mathematical paradigm. Some cosmologists believe in an "open" model, a sparsely dense cosmos that yields a saddle-shaped field that will ultimately end in entropy death as its expansion relentlessly decreases the density of energy and matter. Others favor a more critically dense "closed" model shaped like a ball which will ultimately end with a collapse back into singularity when its elements eventually lose inertia to gravity. Most cosmologists embrace a flat universe. They contend time is a "fourth dimension". In "spacetime", X-Y-Z Euclidean axes aren't linear and perpendicular but curve back upon themselves in a Riemannian topography. The cosmos is still spherical but it's expanding like a bubble and it, too, will experience entropy death as the bubble grows forever. That flat-but-folded universe is deemed to be finite but unbounded inasmuch as one may travel the finite surface of an expanding orb eternally without ever reaching an end. Throwing my Euclidean world a Riemannian curve will require a bit more than just speculative equations that look good on paper.

 In a finite universe, for every point 'A' there would exist some point 'B' within a finite measure at which motion in any direction would not add to the distance between them. If you ever discover such a place in the cosmos, be sure to let me know.

A finite cosmos would have a finite amount of "stuff": space, mass, and energy. All of those elements would be quantifiable into measurable volumes with defined physical domains. By simply calculating the volumes and configurations one should be able to determine some Euclidean X-Y-Z coordinate where everything, even space, should begin to non-exist.

Some believe space may be infinite but matter is not. WHAT? An infinite expanse of space devoid of matter requires no less logical justification than an infinite expanse of matter devoid of space. Space may be relatively inert, but it occupies volume and dissipates energy and that is VERY important. If space were "nothing", then nothing would exist between Earth and Sol.. Everyone and everything would fry. If you can point to something, it exists. Space isn't material in nature, but it's there.

Scholars are quick to remind us no point of infinity exists and they are absolutely correct. That doesn't mean the universe is finite. Infinity is the non-existence of a limit and if a non-existence existed it wouldn't be a non-existence. There's a finite distance between every two points in the universe, but there is no point, however distant, where it ends. There is no all, there is always more.

If the universe was not created, there is no evidence or tenet of logic that suggests it is finite, but unfortunately it would be way too expensive and time consuming to prove this empirically.

~~

## A la Descartes

"Cogito ergo sum." I think, therefore I am. Certainly one must exist in order to experience, and the fact you experience is convincing proof you exist. You likely consider yourself to be "an existence", else you'd call yourself "we" instead of "I". "An existence" is an entity – an irreducible particle either with or without the property of mass. Everything comprised of those entities is a composite. "An existence" is not a composite and a composite – though it may have a unique label – is not "an existence". Your body is composed of a gazillion independent existences. Each one has a separate location, a unique identity and set of properties. For those of you not familiar with higher math, a gazillion is $1 \times 10^N$ where 'N' would be a string of 9s written in four point Arial Narrow typeface on a paper strip that stretches from the cafeteria at Cambridge University to the most distant object in the Oort Cloud. Never attempt to count to a gazillion. So far seven people have tried. Three of them abandoned the effort after more than a decade; two died of natural causes and failed to complete the task. One was just institutionalized and the other one recently committed suicide when he lost count and had to start over.

But I digress. Your body is a composite. Each irreducible particle preexisted your birth and will ultimately survive your demise. Each has a unique history, a separate location, individual properties and physical domain. Logically this is a conundrum. A collection of particles would have just as many identities as there are elements in the set. How can you be "an existence" if that manifestation which you consider to be yourself is a composite?

To reconcile this disparity, hordes of scholarly pundits with names basking in beakers of alphabet soup profess if you toss just the right combination of terrestrial ingredients into a primordial cauldron and stir it really, really hard for a very, very long time, you can create a composite that thinks, propagates and experiences a single existence with an individual identity. Now that may sound silly (I call it the Pinocchio hypothesis), but which lowly layman in his right mind would dare contradict an entire horde of scholarly pundits – especially while their names are basking in beakers of alphabet soup. So, with an eye of newt and wing of bat, a pinch of this and a dash of that, these pundits explain away their egregious departure from logic by imbuing a common natural phenomenon called emergent properties (EP) with some extra, more magical powers which prompt them to cite biochemical evolution as the sole source of all life on Earth.

But even the most tenured of scholars aren't able to explain the specific mechanics of EP that transform a body with $8 \times 10^{27}$ atoms into a single existence with an individual identity. In fact, there seems to be two distinct factions in the EP camp. The "integration" group assures us some mysterious power of unification melds a composite into a single identity and awareness. They tell us $8 \times 10^{27} = 1$. On the other hand, the "emergence" group tries to convince us $8 \times 10^{27} = 8 \times 10^{27} + 1$, claiming any sense of self is due to the whole being greater than the sum of its parts. They expect us to believe composites can conjure up a supervening entity, a temporary ego or virtual being with its own separate awareness and identity. In their practice of this mathematical sorcery, proponents of EP have been idiomatically reduced to casting the incantations "integrated" and "emergent" because both "abracadabra" and "hocus pocus" are currently shunned and disfavored by the orthodox scientific community.

Hogwarts! If that is science, then Harry Potter is the next Isaac Newton. If you believe you are the product of emergent properties you're claiming to be an occurrence and not an existence. Merlin, himself, would have been embarrassed by such magical thinking. To quote Sir Arthur Conan Doyle's famous character Sherlock Holmes in Chapter six of *The Sign of Four*: "when you've eliminated the impossible, whatever remains, however improbable, must be the truth." My high school geometry teacher warned me W.E.C.I.B. (What Else Could It Be) is not a viable postulate, but employing it suggests life is simply the product of a spectrum of yet undiscovered entities with the attribute of natural animation who long ago began to manipulate the resources of this planet or "wear the mud" so to speak. Some call them "souls"; others call them "spirits" or "life forces". Regardless what you call them, they are the impetus from which all forms of life arise.

Our physical size is extremely tiny prior to our trek into life (a feature for which anyone who is, was, or ever will become pregnant can be eternally grateful), so it comes as no surprise we haven't been able to isolate and identify that element within that compiled and now compels our corporal garb. Your body is something you wear not something you are. It does; however, seem to be a necessary tool in order for us to function and think in human terms. By rote and repetition you've been trained since birth to believe you are what you see in the mirror – hair, eyes, nose, skin, and appendages. You developed a self-image that your body is YOU. But your corpse is entirely removable; demonstrably so. If you cut off an appendage, it will suddenly be "over there", yet you will not lose your identity. You may even retain feeling in a phantom limb that isn't there. Just because something was held onto your body by molecular bond didn't make it YOU. Your body is simply the remnants of that cheeseburger and fries you ate a few years ago, that beer you had yesterday and that delicious Cesar Salad from 2010. Most of the cells you wear today will be replaced by new cuisine within the next seven years or so. The brain is said to be the home of the id, yet you can remove almost half of it (a requirement to run for public office) and still retain your identity. There is not a single cell in your entire body whose removal would cause you significant discomfort, much less destroy your sense of self.

As strange as it may seem, you have no idea what you actually look like because consciousness, as we know it, only occurs when you are wrapped within your corporal shell. Even if you could strip away the blood and the bones just long enough to glimpse your true countenance, you might see nothing at all, for that fundamental element which is you may not have the property of mass. It may be infinitesimal and like space, your essence may be transparent – more invisible than the air you breathe.

Where did life come from? It came from everywhere. You, I and every living thing on this planet were elsewhere fifty trillion years ago.

Are all life forces fungible? Am I human just due to the luck of the draw? Could I have incarnated as a microbe, a tree or an alligator just as easily?

If all life forces are fungible, the current structures assumed by Earthly critters must be attributed to either the luck of the draw or the condition of each life force when it entered the realm of our planet. Did trees come from K-Pax? Did snails come from Zog, Planet-X or one of the 75 planets ruled by Xenu? If there are specific kinds of life forces on Earth, how many? Evolution may conceal the true number. Were they embedded in the space dust that eventually became Earth or did they arrive later – possibly even recently? Are there other varieties not found on our planet? Again, how many, and will new ones find their way here?

Does life recycle (reincarnation)? Much if not everything in nature is cyclical. You are alive now (I hope) and reincarnation would be no stranger than incarnation. Why can't you remember past lives? Have you ever tried to play a 78 rpm record on a DVD player? You're wearing the most recent upgrade modern corporal engineering can offer. Consciousness and memory seem to be dependent on our mortal coil. Do you miss and long to be reunited with loved ones from lives past (probably not)? Death is a big red reset button. You likely revert back into sub-microscopic form (the more concentrated the more potent), your metabolism slows almost to a halt (years can pass in an instant), and both instinct and the winds of fate eventually find you another host as your death state matures. This is, of course, pure conjecture, but having been dead as recently as 1948, I feel my experience as a decedent makes my conjecture as valid as that of any who care to abuse the subject.

Do you yearn for immortality? That would certainly be foolish. Would you prefer to have a Homo habilis appearance with a commensurate IQ? Would you enjoy crawling around as a silicon-based bug or in that same worm frame you inhabited when a massive white-hot star went extinct trillions of eons ago?

~~

**Conclusion:**

Well, we've made the universe infinite and eternal again and we canceled its expiration date. Please hold your applause. This isn't exactly rocket science; it requires no esoteric equations, no orbiting telescopes or expensive particle accelerators; you don't need a degree in cosmology, math or physics;

or even a high school education to understand it. Fooled by nature's sleight of hand and lost in the complexity of their esoteric equations, Einstein, Hubble and the entire discipline of cosmology seem to have ignored the obvious. Science has as much (if not more) dogma and politics as religion and in the plush, publish-or-perish ivory towers of academia, hypotheses touting multiverses, extra dimensions and cosmic expansion into entropy death are where the real money is (some strings attached). Beautiful equations can describe fantasy as easily as fact, but without the capacity to parse differentials with any degree of integrity, how could any lowly layman dare challenge the sanity of such sophisticated scholarly branes?

    You didn't become immortal but you did become eternal. Life and death are just physical conditions, transient states of being. Existence is eternal. When you die you will be dead - but you will still be. Be kind to animals, they may be old relatives.

There will come a time in the history of mankind when future societies will look back upon our modern era and wonder how creatures who couldn't even understand the nature of their own being could have considered themselves intelligent when the evidence that surrounded them was so obvious and compelling.

How was the universe created? When did it begin? Falsely premised questions have no answers. I hope you enjoyed the read. Once it sinks in, you'll see it poses many more questions than answers. But at least they will be the right questions.

www.ingramcontent.com/pod-product-compliance
Lightning Source LLC
Chambersburg PA
CBHW040055250526
45473CB00041B/500